Pollinator Garden

Your Perfect Companion for a Beautiful Pollinator Garden

Create a vibrant garden buzzing with bees, butterflies, hummingbirds, and other beneficial pollinators.

ABOUT THE AUTHOR

Lisa Monias is a designer, illustrator, and lifelong nature enthusiast. She enjoys gardening, drawing, and creating books that inspire curiosity about the natural world. She hopes this guide encourages readers to plant more flowers, support pollinators, and enjoy the beauty of every season.

Designed by:
Lisa Monias
Printed in the
United States of America

Hardcover ISBN: 979-8-950843-04-4
Paperback ISBN: 979-8-950843-05-1

Art and images licensed
from Shutterstock.

Find Your Perfect Plants

Welcome to your Pocket Pollinator Garden! This guide makes it easy to discover beautiful plants that help pollinators thrive. Each plant profile includes a brief description, bloom season, growing conditions, USDA hardiness zone, and the pollinators it attracts.

Use the bloom calendar bar to choose plants that provide nectar and pollen throughout the growing season. Mix flowers, herbs, trees, and shrubs with overlapping bloom times to create a continuous food source for bees, butterflies, hummingbirds, moths, and other beneficial wildlife.

Whether you're planting a garden, adding a few containers, or simply exploring nature, this guide is designed to help you create a landscape that's both beautiful and welcoming to pollinators.

CONTENTS

ANNUALS

PERENNIALS

HERBS

TREES & SHRUBS

Growing Zones

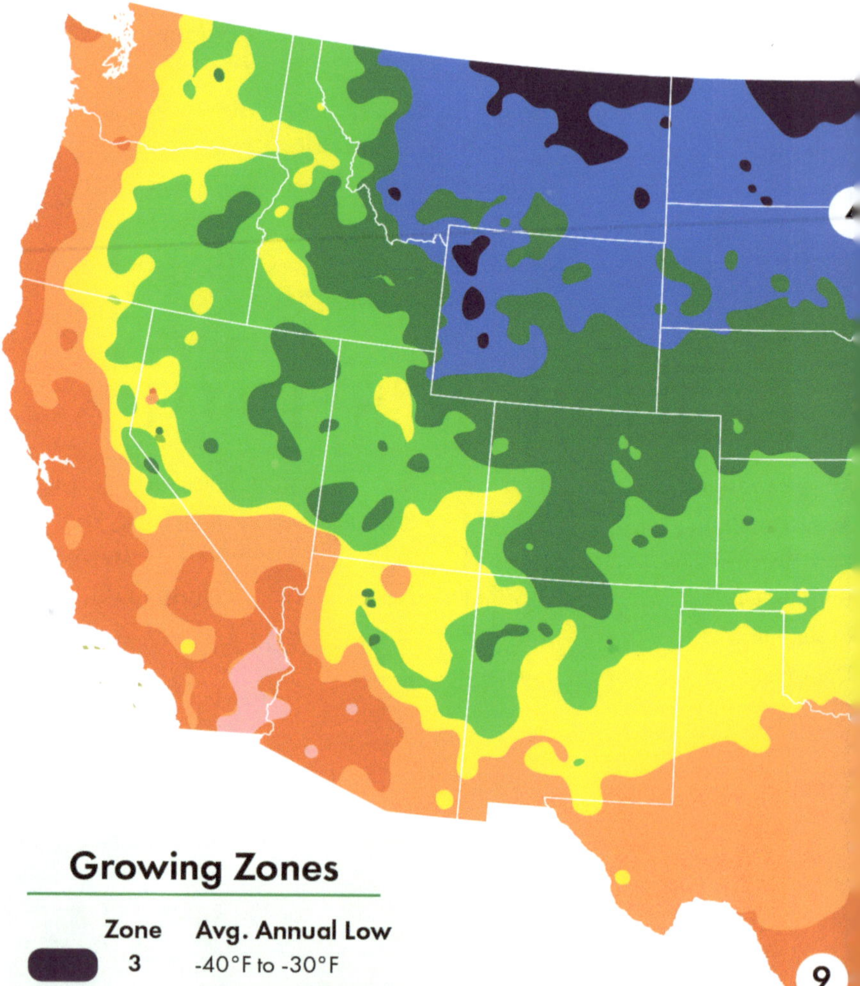

Growing Zones

Zone	Avg. Annual Low
3	-40°F to -30°F
4	-30°F to -20°F
5	-20°F to -10°F
6	-10°F to 0°F
7	0°F to 10°F
8	10°F to 20°F
9	20°F to 30°F
10	30°F to 40°F

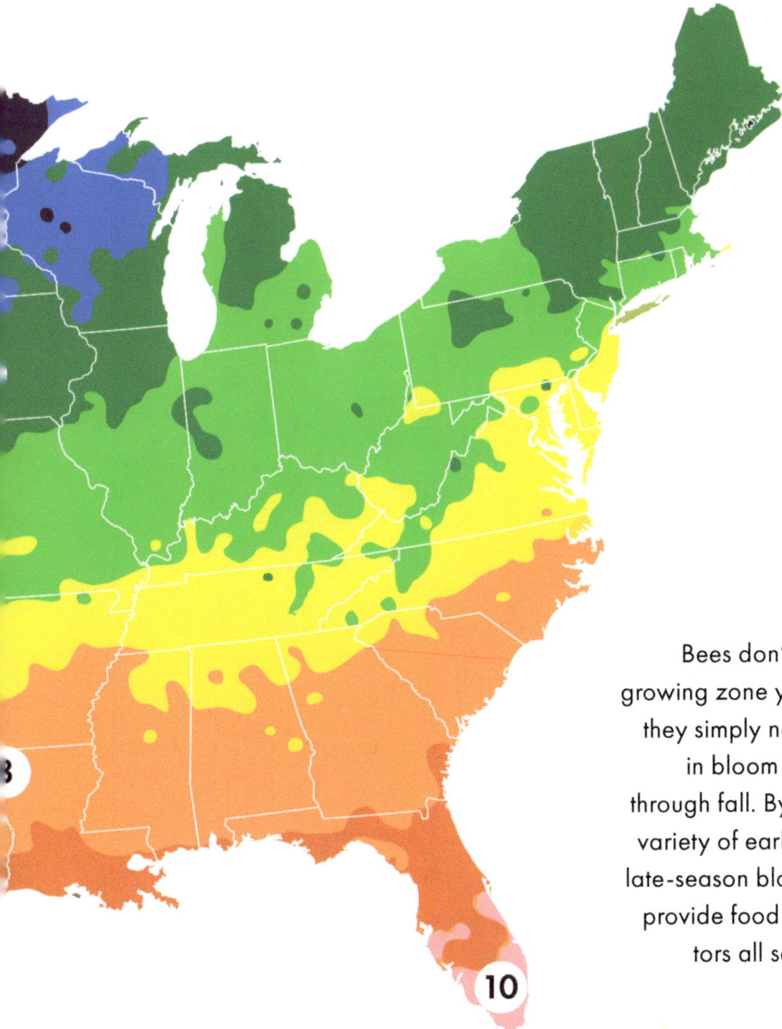

Bees don't care what growing zone you live in—they simply need flowers in bloom from spring through fall. By planting a variety of early, mid, and late-season blooms, you'll provide food for pollinators all season long.

How to Use Growing Zones

- Your growing zone is based on your area's average coldest winter temperature.
- Choose plants that match or are hardy to your zone.
- Local sunlight, rainfall, and soil are just as important as your growing zone.
- When in doubt, ask your local nursery for the best varieties.

CREATE A GARDEN

Pollinator Paradise

Pollinators do far more than add beauty to the garden—they help produce many of the fruits, vegetables, herbs, and wildflowers we enjoy every day. By planting a variety of nectar-rich flowers that bloom from spring through fall, providing host plants, fresh water, and avoiding harmful pesticides, you can create a welcoming habitat that supports bees, butterflies, hummingbirds, moths, and many other beneficial insects throughout the seasons.

> Plant flowers that bloom from spring through fall. Group similar plants together so pollinators can find them more easily.

Even a small backyard, patio, or balcony can make a difference. Every flower planted becomes another source of food, and every bloom is another stop along a pollinator's journey. Together, thousands of small gardens create healthy habitats that help these remarkable creatures thrive for generations to come.

Grow a Garden That Buzzes with Life

FLOWERS
- Coneflower
- Bee Balm
- Goldenrod
- Yarrow

HERBS
- Basil
- Dill
- Fennel
- Thyme

TREES & SHRUBS
- Buttonbush
- Redbud
- Serviceberry
- Sourwood
- American Holly

Pollinators You'll Welcome

Bees – Nature's most important pollinators.

Butterflies – Add beauty while pollinating flowers.

Hummingbirds – Visit tubular blooms for nectar.

Moths – Important nighttime pollinators.

Beneficial Insects – Ladybugs, lacewings, hoverflies, praying mantises, and other helpful insects help gardens stay healthy.

CREATE A GARDEN FOR

Monarch Butterflies

Every year, millions of monarch butterflies make one of the longest migrations in the insect world, traveling thousands of miles across North America. Along the way, they depend on healthy habitats filled with nectar-rich flowers and milkweed. By planting a monarch-friendly garden, you can provide food for both adult butterflies and their caterpillars while helping support future generations.

> Provide a shallow dish of water with small stones for butterflies to safely drink.

Watching monarchs flutter through your garden is one of the greatest rewards of gardening. Their brilliant orange wings bring beauty and movement to every landscape, while their remarkable life cycle—from tiny egg to striped caterpillar, emerald chrysalis, and beautiful butterfly—is fascinating for both children and adults. Even a small garden can make a meaningful difference.

Favorite nectar plants for adult butterflies

- Aster
- Bee Balm
- Blazing Star (Liatris)
- Black-Eyed Susan
- Blanket Flower
- Coneflower
- Coreopsis
- Garden Phlox
- Goldenrod
- Joe-Pye Weed
- Lantana
- Lavender
- Penstemon
- Salvia
- Sedum 'Autumn Joy'
- Shasta Daisy
- Verbena
- Yarrow
- Zinnia

Host plants where monarchs lay their eggs and caterpillars feed

- Butterfly Weed
- Swamp Milkweed
- Common Milkweed
- Showy Milkweed
- Tropical Milkweed*

*In warm southern climates, many experts recommend cutting tropical milkweed back in late fall or choosing native milkweed species instead.

Annuals

SOIL MOISTURE: Medium

SUN: Full Sun to Part Sun (4–8+ hours)

PLANT: Spring the last frost

ZONES: 2–11 (Annual)

Ageratum

Ageratum houstonianum

Ageratum is a compact annual that produces fluffy blue, purple, pink, or white flower clusters from late spring until frost. Its long-lasting blooms attract bees, butterflies, and other beneficial pollinators, making it an excellent choice for borders, containers, and flower beds.

GROWING TIP:
Deadhead spent flowers regularly to encourage continuous blooming.

POLLINATOR POWER: ★★★★★

COMPANION PLANTS:
Marigold • Zinnia • Salvia
Petunia • Verbena

HEIGHT: 6–24 in. **SPACING:** 6–12 in.

NOURISHES:

BEES BUTTERFLIES MOTHS

Its nectar-rich flowers attract a variety of pollinators while adding long-lasting color to borders, containers, and hanging baskets.

FUN FACT:
The name Ageratum comes from the Greek word meaning "ageless," because the flowers stay fresh and colorful for an exceptionally long time.

JAN	FEB	MAR	APR	MAY	JUN	JUL	AUG	SEP	OCT	NOV	DEC

SOIL MOISTURE: Medium

SUN: Full Sun (6–8+ hours)

PLANT: Early Spring after the last frost

ZONES: 2–11 (Annual)

Bachelor's Button

Centaurea cyanus

Bachelor's Button is an easy-to-grow annual with cheerful blue, pink, purple, white, or burgundy blooms that brighten the garden from late spring through summer. Its nectar-rich flowers attract bees, butterflies, and other beneficial pollinators while making excellent cut flowers.

GROWING TIP:
Direct sow seeds where they will grow, since Bachelor's Button does not transplant well.

POLLINATOR POWER: ★★★★

COMPANION PLANTS:
Cosmos • Zinnia • Coreopsis
Calendula • Larkspur • Poppy

HEIGHT: 12–36 in. **SPACING:** 6–12 in.

NOURISHES:

BEES BUTTERFLIES BENEFICIAL INSECTS

Its nectar-rich blooms provide an excellent food source for pollinators while adding vibrant color to gardens and flower beds.

FUN FACT:
Young men once wore the flowers in their lapels to show they were in love. If the bloom faded quickly, it was said their love was not returned.

JAN	FEB	MAR	APR	MAY	JUN	JUL	AUG	SEP	OCT	NOV	DEC

SOIL MOISTURE: Medium

SUN: Full Sun (6–8+ hours)

PLANT: After the last frost or start indoors 4–6 weeks early

ZONES: 2–11

Borage

Borago officinalis

Borage is a fast-growing annual with clusters of brilliant star-shaped blue flowers that bloom from late spring through fall. Its nectar-rich blossoms are beloved by bees, butterflies, and other beneficial insects, making it one of the best flowers for supporting pollinators in the garden.

GROWING TIP:
Give borage plenty of room to spread. It readily self-seeds, providing beautiful blooms year after year.

POLLINATOR POWER: ★★★★★

COMPANION PLANTS:
Tomatoes • Strawberries • Squash
Cucumbers • Beans • Marigolds

HEIGHT: 18–36 in. **SPACING:** 12–18 in.

NOURISHES:

BEES · BUTTERFLIES · HUMMINGBIRDS · MOTHS

Provides abundant nectar for bees, butterflies, hummingbirds, and many other pollinators.

FUN FACT:
Borage has long been known as the "Bee Plant" because honeybees visit its flowers from morning until evening. A single plant can produce hundreds of blooms.

JAN	FEB	MAR	APR	MAY	JUN	JUL	AUG	SEP	OCT	NOV	DEC

SOIL MOISTURE: Average | **SUN:** Full Sun to Part Sun | **PLANT:** Early spring or late summer/fall | **ZONES:** 2–11

Calendula

Calendula officinalis

Calendula is a cheerful, easy-to-grow annual prized for its sunny yellow and orange blooms. Flowering from spring through fall, it attracts bees, butterflies, and other beneficial insects while adding long-lasting color to flower beds, vegetable gardens, borders, and containers.

GROWING TIP:
Grows well in containers and often reseeds itself for next year.

POLLINATOR POWER: ★★★★★

COMPANION PLANTS:
Calendula • Cosmos • Nasturtium
Sunflower • Zinnia

HEIGHT: 12–24 in. **SPACING:** 8–12 in.

NOURISHES:

BEES BUTTERFLIES

Gardeners love Calendula because it attracts bees, butterflies, hoverflies, and other beneficial insects.

FUN FACT:
Calendula petals are edible and are sometimes called "poor man's saffron." Their bright petals have been used for centuries to add color to salads, soups, and herbal teas.

JAN	FEB	MAR	APR	MAY	JUN	JUL	AUG	SEP	OCT	NOV	DEC

SOIL MOISTURE: Medium

SUN: Full Sun (6–8+ hours)

PLANT: After the last frost or indoors 6–8 weeks

ZONES: 2–11

Cleome

Cleome hassleriana

Cleome is a tall, eye-catching annual with airy clusters of pink, white, purple, or lavender flowers that resemble colorful fireworks. Blooming from early summer until frost, its nectar-rich blossoms attract bees, butterflies, hummingbirds, and other beneficial insects while adding height and elegance to flower beds.

POLLINATOR POWER: ★★★★★

COMPANION PLANTS:

Cosmos • Zinnias • Sunflowers
Salvia • Verbena • Marigolds

HEIGHT: 36–60 in. **SPACING:** 12–18 in.

NOURISHES:

BEES BUTTERFLIES HUMMINGBIRDS MOTHS

GROWING TIP:

Once established, Cleome tolerates heat, drought, and often self-seeds.

FUN FACT:

Cleome is commonly called "Spider Flower" because its long, graceful stamens give each bloom the appearance of a colorful spider with delicate legs.

JAN	FEB	MAR	APR	MAY	JUN	JUL	AUG	SEP	OCT	NOV	DEC

SOIL MOISTURE: Dry to Medium

SUN: Full Sun (6–8+ hours)

PLANT: After the last frost

ZONES: 2–11

Coreopsis (Plains)

Coreopsis tinctoria

Plains Coreopsis is a cheerful native annual with daisy-like flowers in brilliant shades of golden yellow and deep red. Blooming from early summer through fall, its nectar-rich blossoms attract bees, butterflies, and other beneficial pollinators while bringing long-lasting color.

GROWING TIP:
Direct sow seeds after the last frost in a sunny location. Plains Coreopsis thrives in poor soils and is drought tolerant.

POLLINATOR POWER: ★★★★★

COMPANION PLANTS:
Blanket Flower • Black-Eyed Susan
Cosmos • Coneflower • Bee Balm
Purple Prairie Clover

HEIGHT: 24–48 in. **SPACING:** 12–18 in.

NOURISHES:

BEES BUTTERFLIES HUMMINGBIRDS MOTHS

It provides an excellent food source for bees, butterflies, and other pollinators.

FUN FACT:
Plains Coreopsis is sometimes called "Golden Tickseed" because its tiny seeds resemble ticks. Despite the unusual nickname, it's treasured for its bright blooms.

JAN	FEB	MAR	APR	MAY	JUN	JUL	AUG	SEP	OCT	NOV	DEC

SOIL MOISTURE: Low to Medium

SUN: Full Sun (6–8+ hours)

PLANT: Spring after frost

ZONES: 2–11

Cosmos

Cosmos bipinnatus

Cosmos is a graceful annual with delicate, fern-like foliage and daisy-like flowers in shades of pink, white, crimson, and lavender. Blooming from early summer until frost, its nectar-rich blossoms attract bees, butterflies, and other beneficial pollinators while adding airy beauty and vibrant color.

GROWING TIP:
Deadhead spent blooms for more flowers.

POLLINATOR POWER: ★★★★★

COMPANION PLANTS:
Zinnias • Cleome • Marigolds
Bachelor's Button • Sunflowers
Coreopsis

HEIGHT: 24–60 in. **SPACING:** 12–18 in.

NOURISHES:

BEES BUTTERFLIES HUMMINGBIRDS MOTHS

Its nectar-rich blooms provide a dependable food source for beneficial pollinators.

FUN FACT:
Cosmos often blooms more abundantly in average or poor soil. Too much fertilizer encourages leafy growth with fewer flowers.

JAN	FEB	MAR	APR	MAY	JUN	JUL	AUG	SEP	OCT	NOV	DEC

SOIL MOISTURE: Low to Medium

SUN: Full Sun (6–8+ hrs)

PLANT: Spring after the last frost

ZONES: 2–11

Globe Amaranth

Gomphrena globosa

Globe Amaranth is a heat-loving annual with cheerful, clover-like blooms in shades of purple, pink, white, red, and orange. Blooming from early summer until frost, its long-lasting blooms attract bees, butterflies, and other beneficial pollinators while adding vibrant color to gardens and containers.

GROWING TIP:
Remove faded blooms for more flowers.

POLLINATOR POWER: ★★★★★

COMPANION PLANTS:
Zinnias • Cosmos • Salvia
Marigolds • Celosia • Sunflowers

HEIGHT: 12–24 in. **SPACING:** 9–12 in.

NOURISHES:
BEES · BUTTERFLIES · HUMMINGBIRDS · MOTHS

Its nectar-rich blooms provide a dependable food source for bees, butterflies, and other beneficial pollinators.

FUN FACT:
Globe Amaranth flowers retain their bright color when dried, making them one of the best flowers for dried arrangements and wreaths.

JAN	FEB	MAR	APR	MAY	JUN	JUL	AUG	SEP	OCT	NOV	DEC

SOIL MOISTURE: Medium

SUN: Full Sun to Partial Shade (4–8+ hours)

PLANT: Spring after the last frost

ZONES: 2–11

Lobelia

Lobelia erinus

Lobelia is a compact annual with masses of brilliant blue, purple, white, or pink flowers that bloom from spring through fall. Its nectar-rich blossoms attract bees, butterflies, hummingbirds, and other beneficial pollinators while adding vibrant color to borders, containers, and hanging baskets.

GROWING TIP:
Trim plants lightly after the first bloom for more flowers.

POLLINATOR POWER: ★★★★★

COMPANION PLANTS:
Alyssum • Petunias • Pansies
Bacopa • Verbena • Sweet Potato Vine

HEIGHT: 6–12 in. **SPACING:** 6–8 in.

NOURISHES:

BEES BUTTERFLIES HUMMINGBIRDS MOTHS

Its nectar-rich blooms provide a dependable food source for bees, butterflies, hummingbirds, and other beneficial pollinators.

FUN FACT:
Lobelia blooms most heavily during cool weather and often flowers again in fall as temperatures become milder.

JAN	FEB	MAR	APR	MAY	JUN	JUL	AUG	SEP	OCT	NOV	DEC

SOIL MOISTURE: Low to Medium

SUN: Full Sun (6–8+ hours)

PLANT: Spring after the last frost

ZONES: 2–11

Marigold

Tagetes spp.

Marigolds are cheerful annuals with bright yellow, orange, red, or golden blooms that flower from late spring until frost. Their nectar-rich blossoms attract bees, butterflies, and other beneficial pollinators while adding vibrant color to gardens, borders, and containers.

GROWING TIP:

Remove faded blooms for more flowers.

POLLINATOR POWER: ★★★★★

COMPANION PLANTS:

Tomatoes • Basil • Zinnias
Cosmos • Nasturtiums • Calendula

HEIGHT: 6–36 in. **SPACING:** 8–12 in.

NOURISHES:

BEES BUTTERFLIES HUMMINGBIRDS MOTHS

Its nectar-rich blooms provide a dependable food source for bees, butterflies, and other beneficial pollinators.

FUN FACT:

Marigolds are famous companion plants because their strong scent may help discourage certain garden pests. Popular varieties include French, African, and Signet Marigolds.

JAN	FEB	MAR	APR	MAY	JUN	JUL	AUG	SEP	OCT	NOV	DEC

SOIL MOISTURE: Low to Medium

SUN: Full Sun (6-8+ hours)

PLANT: Spring after the last frost

ZONES: 2–11

Mexican Sunflower

Tithonia rotundifolia

Mexican Sunflower is a bold annual with brilliant orange to fiery red-orange blooms that flower from midsummer until frost. Its nectar-rich blossoms attract bees, butterflies, hummingbirds, and other beneficial pollinators while adding vibrant color and height to gardens and borders.

GROWING TIP:
Plant in full sun for the most abundant blooms.

POLLINATOR POWER: ★★★★★

COMPANION PLANTS:
Zinnias • Cosmos • Sunflowers
Salvia • Cleome • Verbena

HEIGHT: 48–72 in. **SPACING:** 18–24 in.

NOURISHES:

BEES BUTTERFLIES HUMMINGBIRDS MOTHS

Provides nectar for butterflies, bees, hummingbirds, and other pollinators but is not a host plant for most butterfly species.

FUN FACT:
Mexican Sunflower is a favorite of monarch butterflies during their fall migration because it provides abundant late-season nectar.

JAN	FEB	MAR	APR	MAY	JUN	JUL	AUG	SEP	OCT	NOV	DEC

SOIL MOISTURE: Medium | SUN: Full Sun to Partial Shade (4–8+ hours) | PLANT: Spring after the last frost | ZONES: 2–11

Nasturtium

Tropaeolum majus

Nasturtium is a fast-growing annual with vibrant trumpet-shaped flowers in shades of red, orange, yellow, cream, and salmon. Blooming from late spring until frost, its nectar-rich blossoms attract bees, butterflies, hummingbirds, and other beneficial pollinators while adding cheerful color to gardens, containers, and hanging baskets.

GROWING TIP:
Avoid rich soil to encourage more blooms.

POLLINATOR POWER: ★★★★★

COMPANION PLANTS:
Tomatoes • Cucumbers • Squash
Beans • Marigolds • Radishes

HEIGHT: 12–18 in. **SPACING:** 10–12 in.

NOURISHES:
BEES BUTTERFLIES HUMMINGBIRDS MOTHS

Its nectar-rich blooms provide a dependable food source for bees, butterflies, hummingbirds, and other beneficial pollinators.

FUN FACT:
Both the flowers and leaves are edible, adding a peppery flavor to salads and garnishes.

JAN	FEB	MAR	APR	MAY	JUN	JUL	AUG	SEP	OCT	NOV	DEC

SOIL MOISTURE: Medium

SUN: Full Sun (6-8+ hours)

PLANT: Spring after the last frost

ZONES: 2-11

Petunia

Petunia × atkinsiana

Petunias are colorful annuals with trumpet-shaped blooms in nearly every shade, including pink, purple, red, white, yellow, and bi-colors. Blooming from late spring until frost, their nectar-rich flowers attract bees, butterflies, hummingbirds, and other beneficial pollinators while adding continuous color to gardens, containers, and hanging baskets.

GROWING TIP:

Remove faded blooms for more flowers.

POLLINATOR POWER: ★★★★★

COMPANION PLANTS:

Lobelia • Alyssum • Verbena

Marigolds • Calibrachoa • Dusty Miller

HEIGHT: 6–18 in. **SPACING:** 10–12 in.

NOURISHES:

BEES BUTTERFLIES HUMMINGBIRDS MOTHS

Its nectar-rich blooms provide a dependable food source for bees, butterflies, hummingbirds, moths, and other beneficial pollinators.

FUN FACT:

Petunias release a sweet fragrance in the evening, helping attract night pollinators like moths.

JAN	FEB	MAR	APR	MAY	JUN	JUL	AUG	SEP	OCT	NOV	DEC

SOIL MOISTURE: Low to Medium

SUN: Full Sun (6-8+ hours)

PLANT: Spring after frost

ZONES: 2-11 (Annual)

Phacelia

Phacelia tanacetifolia

Phacelia is a fast-growing annual with delicate, fern-like foliage and unique lavender-blue flowers that unfurl in graceful, curled clusters. Blooming from late spring through summer, its nectar-rich blossoms attract bees, butterflies, and other beneficial pollinators while adding soft color and texture to gardens and wildflower plantings.

GROWING TIP:
Sow after the last frost for best results.

POLLINATOR POWER: ★★★★★

COMPANION PLANTS:
California Poppy • Cosmos
Coreopsis • Bachelor's Button
Yarrow • Clover

HEIGHT: 24–48 in. **SPACING:** 12–18 in.

NOURISHES:

BEES BUTTERFLIES HUMMINGBIRDS MOTHS

Provides nectar for bees, butterflies, and other beneficial pollinators.

FUN FACT:
Often called "Bee's Friend," Phacelia is one of the world's most popular flowers for attracting honeybees.

JAN	FEB	MAR	APR	MAY	JUN	JUL	AUG	SEP	OCT	NOV	DEC

SOIL MOISTURE: Medium

SUN: Full Sun to Partial Shade (4–8+ hours)

PLANT: Spring after the last frost

ZONE: 2–11 (Annual)

Salvia (Scarlet)

Salvia splendens

Scarlet Salvia is a vibrant annual with brilliant red flower spikes that bloom from late spring until frost. Its nectar-rich blossoms attract hummingbirds, bees, butterflies, and other beneficial pollinators while adding bold color and vertical interest to gardens, borders, and containers.

GROWING TIP:

Remove spent flower spikes for more blooms.

POLLINATOR POWER: ★★★★★

COMPANION PLANTS:

Petunias • Marigolds • Zinnias
Lobelia • Alyssum • Coleus

HEIGHT: 12–24 in. **SPACING:** 10–12 in.

NOURISHES:

BEES BUTTERFLIES HUMMINGBIRDS MOTHS

Its nectar-rich blooms provide a dependable food source for hummingbirds, bees, butterflies, moths, and other beneficial pollinators.

FUN FACT:

Scarlet Salvia's tubular flowers are perfectly shaped for hummingbirds, making it one of their favorite garden flowers.

JAN	FEB	MAR	APR	MAY	JUN	JUL	AUG	SEP	OCT	NOV	DEC

SOIL MOISTURE: Medium

SUN: Full Sun to Partial Shade (4–8+ hours)

PLANT: Spring after the last frost

ZONES: 2–11

Snapdragon

Antirrhinum majus

Snapdragons are colorful annuals with tall spikes of dragon-shaped flowers in shades of red, pink, yellow, orange, white, and purple. Blooming from spring through fall, their nectar-rich blossoms attract bees, butterflies, hummingbirds, and other beneficial pollinators while adding bold color and vertical interest to gardens, borders, and containers.

GROWING TIP:
Remove spent flower spikes.

POLLINATOR POWER: ★★★★★

COMPANION PLANTS:
Pansies • Alyssum • Lobelia
Petunias • Dianthus • Calendula

HEIGHT: 12–36 in. **SPACING:** 8–12 in.

NOURISHES:

BEES BUTTERFLIES HUMMINGBIRDS MOTHS

Its nectar-rich blossoms provide a dependable food source for bees, butterflies, hummingbirds, moths, and other beneficial pollinators.

FUN FACT:
Pinch the sides of a Snapdragon flower, and it opens like a tiny dragon's mouth—giving the plant its name.

JAN	FEB	MAR	APR	MAY	JUN	JUL	AUG	SEP	OCT	NOV	DEC

ANNUALS

SOIL MOISTURE: Medium

SUN: Full Sun (6–8+ hours)

PLANT: Spring after the last frost

ZONES: 2–11

Sunflower

Helianthus annuus

Sunflowers are tall, cheerful annuals with large golden blooms that follow the sun as young plants grow. Blooming from midsummer through fall, their nectar-rich flowers attract bees, butterflies, and other beneficial pollinators while their seeds provide food for birds and wildlife.

GROWING TIP:
Plant in full sun and provide support for taller varieties if needed.

POLLINATOR POWER: ★★★★★

COMPANION PLANTS:
Zinnias • Cosmos • Mexican Sunflower • Marigolds • Nasturtiums

HEIGHT: 36–120 in. **SPACING:** 12–24 in.

NOURISHES:
BEES · BUTTERFLIES · HUMMINGBIRDS · MOTHS

Its nectar-rich blooms provide food for bees, butterflies, and other beneficial pollinators, while its seeds nourish birds and wildlife.

FUN FACT:
A single sunflower head can produce hundreds to over a thousand seeds, providing food for birds and other wildlife.

JAN	FEB	MAR	APR	MAY	JUN	JUL	AUG	SEP	OCT	NOV	DEC

SOIL MOISTURE: Medium

SUN: Full Sun to Partial Shade (4–8+ hours)

PLANT: Spring after the last frost

ZONES: 2–11

Sweet Alyssum

Lobularia maritima

Sweet Alyssum is a low-growing annual with fragrant clusters of tiny flowers in shades of white, pink, lavender, and purple. Blooming from spring through fall, its nectar-rich blossoms attract bees, butterflies, and other beneficial pollinators while creating a beautiful carpet of color in gardens, borders, containers, and hanging baskets.

GROWING TIP:
Trim lightly after blooming.

POLLINATOR POWER: ★★★★★

COMPANION PLANTS:
Petunias • Lobelia • Snapdragons
Marigolds • Verbena • Pansies

HEIGHT: 4–8 in. **SPACING:** 6–8 in.

NOURISHES:

BEES BUTTERFLIES HUMMINGBIRDS MOTHS

Its nectar-rich blossoms provide a dependable food source for bees, butterflies, moths, and other beneficial pollinators.

FUN FACT:
Sweet Alyssum's honey-like fragrance is strongest on warm, sunny days, making it a favorite of both gardeners and pollinators.

JAN	FEB	MAR	APR	MAY	JUN	JUL	AUG	SEP	OCT	NOV	DEC

SOIL MOISTURE: Medium

SUN: Full Sun (6–8+ hours)

PLANT: Spring after the last frost

ZONES: 2–11

Verbena

Verbena × hybrida

Verbena is a long-blooming annual with clusters of colorful flowers in shades of purple, pink, red, white, blue, and bi-colors. Blooming from late spring until frost, its nectar-rich blossoms attract bees, butterflies, hummingbirds, and other beneficial pollinators while adding vibrant color to gardens, borders, containers, and hanging baskets.

GROWING TIP:

Remove spent blooms for more flowers.

POLLINATOR POWER: ★★★★★

COMPANION PLANTS:

Petunias • Sweet Alyssum • Lobelia
Marigolds • Zinnias • Salvia

HEIGHT: 6–18 in. **SPACING:** 12–18 in.

NOURISHES:

BEES BUTTERFLIES HUMMINGBIRDS MOTHS

Its nectar-rich blooms provide a dependable food source for bees, butterflies, hummingbirds, moths, and other beneficial pollinators.

FUN FACT:

Verbena is a favorite nectar plant for butterflies and often blooms continuously until the first frost with regular deadheading.

JAN	FEB	MAR	APR	MAY	JUN	JUL	AUG	SEP	OCT	NOV	DEC

SOIL MOISTURE: Medium

SUN: Full Sun (6–8+ hours)

PLANT: Spring after the last frost

ZONES: 2–11

Zinnia

Zinnia elegans

Zinnias are colorful annuals with bright daisy-like flowers in shades of red, pink, orange, yellow, white, purple, and bi-colors. Blooming from early summer until frost, their nectar-rich blossoms attract bees, butterflies, hummingbirds, and other beneficial pollinators while adding long-lasting color to gardens, borders, and cutting gardens.

GROWING TIP:
Remove spent blooms for more flowers.

POLLINATOR POWER: ★★★★★

COMPANION PLANTS:
Cosmos • Marigolds • Sunflowers
Verbena • Nasturtiums • Salvia

HEIGHT: 12–48 in. **SPACING:** 8–18 in.

NOURISHES:

BEES BUTTERFLIES HUMMINGBIRDS MOTHS

Its nectar-rich blossoms provide food for bees, butterflies, hummingbirds, moths, and other beneficial pollinators.

FUN FACT:
The more you cut Zinnias for bouquets, the more flowers the plants often produce throughout the season.

JAN	FEB	MAR	APR	MAY	JUN	JUL	AUG	SEP	OCT	NOV	DEC

Perennials

PERENNIALS

SOIL MOISTURE: Medium | **SUN:** Full Sun (6–8+ hours) | **PLANT:** Spring after frost | **ZONES:** 3–8

Aster

Symphyotrichum spp.

Asters are hardy perennials with daisy-like flowers in shades of purple, pink, blue, and white that bloom from late summer through fall. Their nectar-rich blossoms attract bees, butterflies, and other beneficial pollinators while providing an important late-season food source.

GROWING TIP:
Divide plants every few years to keep them healthy and blooming well.

POLLINATOR POWER: ★★★★★

COMPANION PLANTS:
Goldenrod • Coneflower
Black-Eyed Susan • Joe-Pye Weed

HEIGHT: 12–48 in. **SPACING:** 12–24 in.

NOURISHES:

BEES BUTTERFLIES BENEFICIAL INSECTS

Its nectar-rich blooms provide a dependable food source for bees, butterflies, moths, and other beneficial pollinators.

FUN FACT:
Asters are among the last flowers to bloom each year, providing valuable nectar for migrating monarch butterflies and late-season bees.

JAN	FEB	MAR	APR	MAY	JUN	JUL	AUG	SEP	OCT	NOV	DEC

SOIL MOISTURE: Medium

SUN: Full Sun to Partial Shade (4–8+ hours)

PLANT: Spring after frost

ZONES: 3–9

Bee Balm

Monarda didyma

Bee Balm is a fragrant perennial with showy blooms in shades of red, pink, purple, and lavender that bloom from early to late summer. Its nectar-rich flowers attract bees, butterflies, hummingbirds, and other beneficial pollinators while adding bold color and fragrance to the garden.

GROWING TIP:
Divide clumps every few years to improve airflow and reduce mildew.

POLLINATOR POWER: ★★★★★

COMPANION PLANTS:
Coneflower • Black-Eyed Susan
Garden Phlox • Salvia • Yarrow

HEIGHT: 24–48 in. **SPACING:** 18–36 in.

NOURISHES:

BEES · BUTTERFLIES · HUMMINGBIRDS · BENEFICIAL INSECTS

Its nectar-rich blooms provide a dependable food source for bees, butterflies, hummingbirds, moths, and other beneficial pollinators.

FUN FACT:
Bee Balm leaves have a minty fragrance and can be brewed into a flavorful herbal tea.

JAN	FEB	MAR	APR	MAY	**JUN**	**JUL**	**AUG**	SEP	OCT	NOV	DEC

SOIL MOISTURE: Low to Medium

SUN: Full Sun (6–8+ hours)

PLANT: Spring after frost

ZONES: 3–9

Black-Eyed Susan

Rudbeckia hirta

Black-Eyed Susan is a cheerful perennial with bright golden-yellow petals surrounding a dark brown center. Blooming from early summer through fall, its nectar-rich flowers attract bees, butterflies, and other beneficial pollinators while adding long-lasting color to gardens and naturalized landscapes.

GROWING TIP:
Deadhead spent blooms to encourage more flowers.

POLLINATOR POWER: ★★★★★

COMPANION PLANTS:
Coneflower • Aster • Bee Balm
Goldenrod • Joe-Pye Weed

HEIGHT: 18–36 in. **SPACING:** 12–18 in.

NOURISHES:

BEES BUTTERFLIES HUMMINGBIRDS MOTHS

Its nectar-rich blooms provide a dependable food source for bees, butterflies, moths, and other beneficial pollinators.

FUN FACT:
Black-Eyed Susans readily self-seed, often returning year after year with even more cheerful blooms.

JAN	FEB	MAR	APR	MAY	JUN	JUL	AUG	SEP	OCT	NOV	DEC

SOIL MOISTURE: Low

SUN: Full Sun (6–8+ hours)

PLANT: Spring after frost

ZONES: 3–10

Blanket Flower

Gaillardia × grandiflora

Blanket Flower is a colorful perennial with daisy-like blooms in brilliant shades of red, orange, and yellow that bloom from early summer through fall. Its nectar-rich blossoms attract bees, butterflies, and other beneficial pollinators while adding vibrant, long-lasting color to gardens.

GROWING TIP:
Deadhead spent blooms to encourage more flowers.

POLLINATOR POWER: ★★★★★

COMPANION PLANTS:
Coneflower • Black-Eyed Susan
Coreopsis • Salvia • Yarrow

HEIGHT: 18–24 in. **SPACING:** 12–18 in.

NOURISHES:

BEES BUTTERFLIES HUMMINGBIRDS MOTHS

Its nectar-rich blooms provide a dependable food source for bees, butterflies, moths, and other beneficial pollinators.

FUN FACT:
Blanket Flower thrives in hot, dry conditions and often continues blooming until the first frost.

JAN	FEB	MAR	APR	MAY	JUN	JUL	AUG	SEP	OCT	NOV	DEC

SOIL MOISTURE: Low to Medium | **SUN:** Full Sun (6–8+ hours) | **PLANT:** Spring after frost | **ZONES:** 3–8

Blazing Star

Liatris spicata

Blazing Star is a striking perennial with tall spikes of fluffy purple flowers that bloom from midsummer through early fall. Its nectar-rich blossoms attract bees, butterflies, hummingbirds, and other beneficial pollinators while adding bold vertical interest to gardens and pollinator beds.

GROWING TIP:
Leave flower stalks standing through fall for added wildlife value.

POLLINATOR POWER: ★★★★★

COMPANION PLANTS:
Coneflower • Black-Eyed Susan
Bee Balm • Aster • Goldenrod

HEIGHT: 24–48 in. **SPACING:** 12–18 in.

NOURISHES:

BEES BUTTERFLIES HUMMINGBIRDS MOTHS

Its nectar-rich blooms provide a dependable food source for bees, butterflies, hummingbirds, moths, and other beneficial pollinators.

FUN FACT:
Blazing Star blooms from the top of the flower spike downward—unlike most flowers, which bloom from the bottom up.

JAN	FEB	MAR	APR	MAY	JUN	JUL	AUG	SEP	OCT	NOV	DEC

SOIL MOISTURE: Low | **SUN:** Full Sun (6-8+ hours) | **PLANT:** Spring after frost | **ZONES:** 3-9

Butterfly Weed

Asclepias tuberosa

Butterfly Weed is a hardy perennial with brilliant clusters of bright orange flowers that bloom from early summer to midsummer. Its nectar-rich blossoms attract bees, butterflies, hummingbirds, and other beneficial pollinators while serving as an important host plant for monarch butterflies.

GROWING TIP:
Avoid transplanting once established because of its deep taproot.

POLLINATOR POWER: ★★★★★

COMPANION PLANTS:
Coneflower • Blazing Star • Black-Eyed Susan • Bee Balm • Aster

HEIGHT: 18-36 in. **SPACING:** 18-24 in.

NOURISHES:

BEES BUTTERFLIES HUMMINGBIRDS MOTHS

Butterfly Weed has nectar-rich blooms and its leaves also serve as an essential host plant for monarch caterpillars.

FUN FACT:
Butterfly Weed is one of the few plants that provides both nectar for adult monarchs and leaves for their caterpillars.

JAN	FEB	MAR	APR	MAY	**JUN**	**JUL**	**AUG**	SEP	OCT	NOV	DEC

PERENNIALS

SOIL MOISTURE: Low

SUN: Full Sun (6–8+ hours)

PLANT: Spring after frost

ZONES: 3–8

Catmint

Nepeta × faassenii

Catmint is a fragrant perennial with soft spikes of lavender-blue flowers that bloom from late spring through early fall. Its nectar-rich blossoms attract bees, butterflies, hummingbirds, and other beneficial pollinators while adding long-lasting color to gardens and borders.

GROWING TIP:
Shear plants lightly after flowering to encourage another flush of blooms.

POLLINATOR POWER: ★★★★★

COMPANION PLANTS:
Salvia • Coneflower • Yarrow
Lavender • Russian Sage

HEIGHT: 12–24 in. **SPACING:** 18–24 in.

NOURISHES:

BEES · BUTTERFLIES · HUMMINGBIRDS · MOTHS

Its nectar-rich blooms provide a dependable food source for bees, butterflies, hummingbirds, moths, and other beneficial pollinators.

FUN FACT:
Catmint is closely related to catnip, but contains much less of the compound that makes many cats go wild.

JAN	FEB	MAR	APR	MAY	JUN	JUL	AUG	SEP	OCT	NOV	DEC

SOIL MOISTURE: Low to Medium

SUN: Full Sun (6–8+ hours)

PLANT: Spring after frost

ZONES: 3–9

Coneflower

Echinacea purpurea

Coneflower is a hardy perennial with large daisy-like blooms in shades of purple, pink, white, and yellow that flower from early summer through fall. Its nectar-rich blossoms attract bees, butterflies, hummingbirds, and other beneficial pollinators while its seed heads provide food for songbirds in fall and winter.

GROWING TIP:
Leave seed heads standing through winter for birds and winter interest.

POLLINATOR POWER: ★★★★★

COMPANION PLANTS:
Black-Eyed Susan • Bee Balm
Blazing Star • Aster • Goldenrod

HEIGHT: 24–48 in. **SPACING:** 18–24 in.

NOURISHES:

BEES BUTTERFLIES BENEFICIAL INSECTS

Its nectar-rich blooms provide a dependable food source for bees, butterflies, hummingbirds, moths, and other pollinators.

FUN FACT:
Coneflower seed heads are a favorite food of American goldfinches during fall and winter.

JAN	FEB	MAR	APR	MAY	JUN	JUL	AUG	SEP	OCT	NOV	DEC

SOIL MOISTURE: Low to Medium

SUN: Full Sun (6–8+ hours)

PLANT: Spring after frost

ZONES: 3–9

Coreopsis

Coreopsis lanceolata

Coreopsis is a cheerful perennial with bright yellow, daisy-like blooms that flower from early summer through fall. Its nectar-rich blossoms attract bees, butterflies, and other beneficial pollinators while adding long-lasting color to gardens, borders, and wildflower plantings.

GROWING TIP:
Deadhead spent blooms to encourage continuous flowering.

POLLINATOR POWER: ★★★★★

COMPANION PLANTS:
Coneflower • Black-Eyed Susan
Blanket Flower • Salvia • Yarrow

HEIGHT: 18–24 in. **SPACING:** 12–18 in.

NOURISHES:

BEES BUTTERFLIES HUMMINGBIRDS MOTHS

Its nectar-rich blooms provide a dependable food source for bees, butterflies, moths, and other beneficial pollinators.

FUN FACT:
Coreopsis is the state wildflower of Florida, where several native species bloom naturally across the state.

JAN	FEB	MAR	APR	MAY	JUN	JUL	AUG	SEP	OCT	NOV	DEC

SOIL MOISTURE: Medium | **SUN:** Full Sun to Partial Shade (4–8+ hours) | **PLANT:** Spring after frost | **ZONES:** 4–8

Garden Phlox

Phlox paniculata

Garden Phlox is a fragrant perennial with large clusters of colorful blooms in shades of pink, purple, white, lavender, and red that flower from midsummer through early fall. Its nectar-rich blossoms attract bees, butterflies, hummingbirds, and other beneficial pollinators while adding bold color and fragrance to gardens.

GROWING TIP:
Provide good air circulation to help prevent powdery mildew.

POLLINATOR POWER: ★★★★★

COMPANION PLANTS:
Coneflower • Bee Balm • Black-Eyed Susan • Aster • Joe-Pye Weed

HEIGHT: 24–48 in. **SPACING:** 18–24 in.

NOURISHES:

BEES BUTTERFLIES HUMMINGBIRDS MOTHS

Its nectar-rich blooms provide a dependable food source for bees, butterflies, hummingbirds, moths, and other beneficial pollinators.

FUN FACT:
Many Garden Phlox varieties produce a sweet fragrance that's especially noticeable on warm summer evenings.

JAN	FEB	MAR	APR	MAY	JUN	JUL	AUG	SEP	OCT	NOV	DEC

SOIL MOISTURE: Low to Medium

SUN: Full Sun (6–8+ hours)

PLANT: Spring after frost

ZONES: 3–9

Goldenrod

Solidago spp.

Goldenrod is a hardy perennial with tall sprays of brilliant golden-yellow flowers that bloom from late summer through fall. Its nectar-rich blossoms attract bees, butterflies, and other beneficial pollinators while providing one of the season's most important late-season food sources.

GROWING TIP:
Divide clumps every few years to keep plants vigorous.

POLLINATOR POWER: ★★★★★

COMPANION PLANTS:
Aster • Coneflower • Joe-Pye Weed
Blazing Star • Black-Eyed Susan

HEIGHT: 24–60 in. **SPACING:** 18–24 in.

NOURISHES:

BEES BUTTERFLIES BENEFICIAL INSECTS

Its nectar-rich blooms provide a dependable late-season food source for bees, butterflies, moths, and other beneficial pollinators.

FUN FACT:
Goldenrod is often mistaken for causing hay fever, but ragweed—not Goldenrod—is usually the real culprit.

JAN	FEB	MAR	APR	MAY	JUN	JUL	AUG	SEP	OCT	NOV	DEC

SOIL MOISTURE: Medium to High

SUN: Full Sun to Partial Shade (4-8+ hours)

PLANT: Spring after frost

ZONES: 4-9

PERENNIALS

Joe-Pye Weed

Eutrochium purpureum

Joe-Pye Weed is a tall perennial with large domed clusters of rosy-pink to mauve flowers that bloom from midsummer through early fall. Its nectar-rich blossoms attract bees, butterflies, hummingbirds, and other beneficial pollinators while adding impressive height and texture to gardens.

GROWING TIP:
Keep soil consistently moist during dry spells.

POLLINATOR POWER: ★★★★★

COMPANION PLANTS:
Aster • Goldenrod • Coneflower
Bee Balm • Garden Phlox

HEIGHT: 48–84 in. **SPACING:** 24–36 in.

NOURISHES:

BEES BUTTERFLIES BENEFICIAL INSECTS

Its nectar-rich blooms provide a dependable food source for bees, butterflies, hummingbirds, moths, and other pollinators.

FUN FACT:
Joe-Pye Weed is named after Joe Pye, a Native American healer who is said to have used the plant for medicinal purposes.

JAN	FEB	MAR	APR	MAY	JUN	JUL	AUG	SEP	OCT	NOV	DEC

SOIL MOISTURE: Low to Medium

SUN: Full Sun (6–8+ hours)

PLANT: Spring after frost

ZONES: 8–11

Lantana

Lantana camara

Lantana is a colorful, easy-to-grow perennial with clusters of bright flowers that bloom for months. Its nectar-rich blossoms attract butterflies, bees, hummingbirds, and other beneficial pollinators while thriving in hot, sunny gardens.

GROWING TIP:
Deadhead spent flower clusters for more blooms.

POLLINATOR POWER: ★★★★★

COMPANION PLANTS:
Salvia • Coneflower • Coreopsis
Bee Balm • Black-Eyed Susan

HEIGHT: 12–72 in. **SPACING:** 18–36 in.

NOURISHES:

BEES BUTTERFLIES HUMMINGBIRDS MOTHS

Provides nectar for butterflies, bees, hummingbirds, and other pollinators. It is not a host plant for most butterfly species.

FUN FACT:
One cluster of Lantana flowers is actually made up of dozens of tiny flowers, and many varieties change color as the blossoms age, yellow flowers may turn orange, pink, or red!

JAN	FEB	MAR	APR	MAY	JUN	JUL	AUG	SEP	OCT	NOV	DEC

SOIL MOISTURE: Low | **SUN:** Full Sun (6-8+ hours) | **PLANT:** Spring after frost | **ZONES:** 5-9

Lavender

Lavandula angustifolia

Lavender is a fragrant perennial with slender spikes of purple flowers that bloom from late spring through summer. Its nectar-rich blossoms attract bees, butterflies, hummingbirds, and other beneficial pollinators while adding beauty, fragrance, and soothing color to gardens.

GROWING TIP:
Avoid overwatering and plant in well-drained soil.

POLLINATOR POWER: ★★★★★

COMPANION PLANTS:
Catmint • Salvia • Coneflower
Russian Sage • Yarrow

HEIGHT: 18-36 in. **SPACING:** 18-24 in.

NOURISHES:

BEES BUTTERFLIES MOTHS

Its nectar-rich blooms provide a dependable food source for bees, butterflies, moths, and other beneficial pollinators.

FUN FACT:
Lavender has been treasured for thousands of years for its calming fragrance and aromatic oils.

JAN	FEB	MAR	APR	MAY	**JUN**	**JUL**	**AUG**	SEP	OCT	NOV	DEC

SOIL MOISTURE: Low to Medium

SUN: Full Sun (6–8+ hours)

PLANT: Spring after frost

ZONES: 3–8

Penstemon

Penstemon digitalis

Penstemon is a hardy perennial with tall spikes of tubular flowers in shades of white, pink, lavender, purple, and red that bloom from late spring through early summer. Its nectar-rich blossoms attract hummingbirds, bees, butterflies, and other beneficial pollinators while adding graceful height to gardens.

GROWING TIP:
Avoid overly wet soil and provide good drainage.

POLLINATOR POWER: ★★★★★

COMPANION PLANTS:
Salvia • Bee Balm • Coneflower
Black-Eyed Susan • Catmint

HEIGHT: 24–48 in. **SPACING:** 18–24 in.

NOURISHES:

BEES BUTTERFLIES HUMMINGBIRDS MOTHS

Its nectar-rich blooms provide a dependable food source for hummingbirds, bees, butterflies, moths, and other beneficial pollinators.

FUN FACT:
The tubular flowers of Penstemon are perfectly shaped for hummingbirds' long beaks and tongues.

JAN	FEB	MAR	APR	MAY	JUN	JUL	AUG	SEP	OCT	NOV	DEC

SOIL MOISTURE: Low | **SUN:** Full Sun (6–8+ hours) | **PLANT:** Spring after frost | **ZONES:** 4–9

Russian Sage

Salvia yangii (formerly Perovskia atriplicifolia)

Russian Sage is a hardy perennial with airy spikes of lavender-blue flowers that bloom from midsummer through fall. Its nectar-rich blossoms attract bees, butterflies, hummingbirds, and other beneficial pollinators while adding soft texture and silvery foliage to gardens.

GROWING TIP:
Avoid overwatering and plant in well-drained soil.

POLLINATOR POWER: ★★★★★

COMPANION PLANTS:
Lavender • Catmint • Salvia
Coneflower • Yarrow

HEIGHT: 36–60 in. **SPACING:** 24–36 in.

NOURISHES:

BEES · BUTTERFLIES · HUMMINGBIRDS · MOTHS

Its nectar-rich blooms provide a dependable food source for bees, butterflies, hummingbirds, moths, and other beneficial pollinators.

FUN FACT:
Russian Sage isn't a true sage—it was reclassified into the same plant genus as garden sages in recent years.

JAN	FEB	MAR	APR	MAY	JUN	**JUL**	**AUG**	**SEP**	OCT	NOV	DEC

SOIL MOISTURE: Low

SUN: Full Sun (6–8+ hours)

PLANT: Spring after frost

ZONES: 4–8

Salvia

Salvia nemorosa

Salvia is a hardy perennial with upright spikes of colorful flowers in shades of purple, blue, pink, white, and red that bloom from late spring into early fall. Its nectar-rich blossoms attract hummingbirds, bees, butterflies, and other beneficial pollinators while adding vibrant color and long-lasting blooms to gardens.

GROWING TIP:

Shear plants after flowering to encourage more blooms.

POLLINATOR POWER: ★★★★★

COMPANION PLANTS:

Lavender • Catmint • Coneflower
Bee Balm • Yarrow

HEIGHT: 18–24 in. **SPACING:** 18–24 in.

NOURISHES:

BEES BUTTERFLIES HUMMINGBIRDS MOTHS

Its nectar-rich blooms provide a dependable food source for hummingbirds, bees, butterflies, moths, and other beneficial pollinators.

FUN FACT:

Many Salvia varieties bloom a second time after being trimmed back following their first flush of flowers.

JAN	FEB	MAR	APR	MAY	JUN	JUL	AUG	SEP	OCT	NOV	DEC

SOIL MOISTURE: Low

SUN: Full Sun (6–8+ hours)

PLANT: Spring after frost

ZONES: 3–9

PERENNIALS

Sedum 'Autumn Joy'

Hylotelephium 'Herbstfreude' (formerly Sedum 'Autumn Joy')

Sedum 'Autumn Joy' or Stonecrop is a hardy perennial with large, flat clusters of rosy-pink flowers that deepen to copper in fall. Its nectar-rich blossoms attract bees, butterflies, and other beneficial pollinators while adding long-lasting color and winter interest to gardens.

GROWING TIP:
Leave flower heads standing through winter for added interest.

POLLINATOR POWER: ★ ★ ★ ★ ★

COMPANION PLANTS:
Aster • Goldenrod • Joe-Pye Weed
Coneflower • Russian Sage

HEIGHT: 18–24 in. **SPACING:** 18–24 in.

NOURISHES:

BEES BUTTERFLIES BENEFICIAL INSECTS

Its nectar-rich blooms provide a dependable late-season food source for bees, butterflies, moths, and other beneficial pollinators.

FUN FACT:
The flower heads change from soft pink to rich copper as the season progresses, providing beauty long after many flowers have faded.

JAN	FEB	MAR	APR	MAY	JUN	JUL	AUG	SEP	OCT	NOV	DEC

SOIL MOISTURE: Low to Medium

SUN: Full Sun (6–8+ hours)

PLANT: Spring after frost

ZONES: 5–9

Shasta Daisy

Leucanthemum × superbum

Shasta Daisy is a cheerful perennial with bright white petals surrounding golden-yellow centers that bloom from early summer through late summer. Its nectar-rich blossoms attract bees, butterflies, and other beneficial pollinators while adding classic beauty and long-lasting blooms to gardens.

GROWING TIP:
Deadhead spent blooms to encourage more flowers.

POLLINATOR POWER: ★★★★★

COMPANION PLANTS:
Coneflower • Black-Eyed Susan
Salvia • Coreopsis • Yarrow

HEIGHT: 24–36 in. **SPACING:** 18–24 in.

NOURISHES:

BEES BUTTERFLIES MOTHS

Its nectar-rich blooms provide a dependable food source for bees, butterflies, moths, and other beneficial pollinators.

FUN FACT:
Shasta Daisy was developed by famous horticulturist Luther Burbank, who named it after the snowy white slopes of Mount Shasta in California.

JAN	FEB	MAR	APR	MAY	JUN	JUL	AUG	SEP	OCT	NOV	DEC

SOIL MOISTURE: Low

SUN: Full Sun (6-8+ hours)

PLANT: Spring after frost

ZONES: 3-9

Yarrow

Achillea millefolium

Yarrow is a hardy perennial with flat clusters of tiny flowers in shades of yellow, white, pink, red, and peach that bloom from early summer through early fall. Its nectar-rich blossoms attract bees, butterflies, and other beneficial pollinators while adding long-lasting color and fine-textured foliage to gardens.

GROWING TIP:
Deadhead spent blooms to encourage more flowers.

POLLINATOR POWER: ★★★★★

COMPANION PLANTS:
Coneflower • Black-Eyed Susan
Salvia • Lavender • Coreopsis

HEIGHT: 18–36 in. **SPACING:** 18–24 in.

NOURISHES:

BEES BUTTERFLIES BENEFICIAL INSECTS

Its nectar-rich blooms provide a dependable food source for bees, butterflies, moths, and other beneficial pollinators.

FUN FACT:
Yarrow is named after the Greek hero Achilles, who according to legend used the plant to treat wounded soldiers.

JAN	FEB	MAR	APR	MAY	JUN	JUL	AUG	SEP	OCT	NOV	DEC

Herbs

Anise Hyssop

Agastache foeniculum

A fragrant herb with lavender-purple flower spikes that attract bees, butterflies, hummingbirds, and other beneficial pollinators. Its sweet, licorice-scented leaves are popular for fresh or dried herbal tea.

NOURISHES:

BEES BUTTERFLIES MOTHS

SUN: Full Sun (6-8+ hrs)

HEIGHT: 24–48 in.

GROWING TIP:
Harvest leaves before flowering for the sweetest flavor, or let plants bloom to attract more pollinators.

GARDEN NOTE:
Fresh or dried leaves make a naturally sweet herbal tea with a mild licorice flavor.

Basil

Ocimum basilicum

A fragrant culinary herb with white flower spikes that attract bees, butterflies, and other beneficial pollinators. Its aromatic leaves are a favorite in gardens and kitchens alike.

NOURISHES:

BEES BUTTERFLIES MOTHS

SUN: Full Sun (6-8+ hrs)

HEIGHT: 12–24 in.

GROWING TIP:
Pinch flower buds for more leafy growth, or allow a few plants to bloom for pollinators.

GARDEN NOTE:
Fresh basil is delicious in pesto, salads, pasta, pizza, and tomato dishes.

Borage

Borago officinalis

A fast-growing herb with brilliant blue, star-shaped flowers that attract bees, butterflies, and other beneficial pollinators. Its edible flowers and cucumber-flavored leaves make it a favorite for pollinator and kitchen gardens.

NOURISHES:

BEES BUTTERFLIES MOTHS

SUN: Full Sun (6–8+ hours)

HEIGHT: 18–36 in.

GROWING TIP:

Allow a few plants to self-seed for beautiful blooms year after year.

GARDEN NOTE:

The edible blue flowers make beautiful garnishes for salads, desserts, and summer drinks.

Catnip

Nepeta cataria

A fragrant herb with small white to pale lavender flowers that attract bees, butterflies, and other beneficial pollinators. Its aromatic leaves delight many cats and can also be brewed into a soothing herbal tea.

NOURISHES:

BEES BUTTERFLIES MOTHS

SUN: Full Sun (6–8+ hours)

HEIGHT: 24–36 in.

QUICK TIP:

Pinch back stems regularly to encourage bushier growth and more flowers.

GARDEN NOTE:

Many cats love rolling, rubbing, and playing in fresh catnip, but not every cat responds to it.

Chives

Allium schoenoprasum

A hardy culinary herb with cheerful purple flower clusters that attract bees, butterflies, and other beneficial pollinators. Its mild onion-flavored leaves and edible blossoms are favorites in both gardens and kitchens.

NOURISHES:

BEES BUTTERFLIES MOTHS

SUN: Full Sun (6–8+ hours)

HEIGHT: 12–18 in.

GROWING TIP:
Harvest leaves regularly to encourage fresh new growth, and allow a few plants to flower for pollinators.

GARDEN NOTE:
The edible purple blossoms add a mild onion flavor and beautiful color to salads, soups, and baked potatoes.

Cilantro

Coriandrum sativum

A flavorful culinary herb with delicate white flower clusters that attract bees, butterflies, and other beneficial pollinators. Its fresh leaves are known as cilantro, while its dried seeds are used as the spice coriander.

NOURISHES:

BEES BUTTERFLIES MOTHS

SUN: Full Sun (6–8+ hours)

HEIGHT: 12–24 in.

GROWING TIP:
Succession plant every few weeks for a continuous harvest, and allow some plants to flower for pollinators.

GARDEN NOTE:
Harvest the fresh leaves as cilantro, then let some plants produce coriander seeds for cooking and baking.

Dill

Anethum graveolens

A feathery culinary herb with bright yellow flower clusters that attract bees, butterflies, and other beneficial pollinators. Its fragrant leaves and seeds are popular for seasoning, while the plant also serves as a host for Black Swallowtail butterfly caterpillars.

NOURISHES:

BEES BUTTERFLIES BENEFICIAL INSECTS

SUN: Full Sun (6–8+ hours)

HEIGHT: 24–48 in.

GROWING TIP:
Succession plant every few weeks for a continuous harvest, and allow some plants to flower for pollinators.

GARDEN NOTE:
Fresh dill leaves flavor seafood, potatoes, salads, and pickles, while the seeds are commonly used for pickling.

Fennel

Foeniculum vulgare

A tall, feathery herb with bright yellow flower clusters that attract bees, butterflies, and other beneficial pollinators. Its aromatic leaves, stems, and seeds are widely used in cooking, while the plant also serves as a host for Black Swallowtail butterfly caterpillars.

NOURISHES:

BEES BUTTERFLIES BENEFICIAL INSECTS

SUN: Full Sun (6–8+ hours)

HEIGHT: 36–72 in.

GROWING TIP:
Allow some plants to flower for pollinators, but remove seed heads if you want to prevent self-seeding.

GARDEN NOTE:
The feathery leaves, crunchy bulb (Florence fennel), and flavorful seeds are all used in a variety of savory dishes.

Hyssop

Hyssopus officinalis

A fragrant herb with upright spikes of blue, purple, pink, or white flowers that attract bees, butterflies, and other beneficial pollinators. Its aromatic leaves have long been valued in herb gardens for both culinary and traditional herbal uses.

NOURISHES:

BEES BUTTERFLIES MOTHS

SUN: Full Sun (6–8+ hours)

HEIGHT: 12–24 in.

GROWING TIP:
Harvest stems before full bloom for the best flavor, or allow plants to flower to attract pollinators.

GARDEN NOTE:
The leaves can be used fresh or dried to flavor teas, soups, and savory dishes.

Lavender

Lavandula angustifolia

A fragrant culinary herb with slender purple flower spikes that attract bees, butterflies, moths, and other beneficial pollinators. Its aromatic flowers and leaves are prized for cooking, teas, crafts, and calming fragrance.

NOURISHES:

BEES BUTTERFLIES MOTHS

SUN: Full Sun (6–8+ hours)

HEIGHT: 18–36 in.

GROWING TIP:
Plant in well-drained soil and avoid overwatering for healthy, long-lived plants.

GARDEN NOTE:
Fresh or dried lavender adds delicate floral flavor to teas, cookies, lemonade, and baked treats.

Lemon Balm

Melissa officinalis

A fragrant herb with small white flowers that attract bees, butterflies, and other beneficial pollinators. Its fresh, lemon-scented leaves are widely enjoyed in teas, desserts, and herbal recipes

NOURISHES:

BEES BUTTERFLIES MOTHS

SUN: Full Sun (6–8+ hours) or Part Shade

HEIGHT: 18–36 in.

GROWING TIP:
Harvest leaves often to encourage fresh growth, and trim plants regularly to keep them tidy.

GARDEN NOTE:
Fresh or dried leaves make a refreshing herbal tea and add bright citrus flavor to salads, fruit, and summer drinks.

Marjoram

Origanum majorana

A sweet, aromatic culinary herb with tiny white to pale pink flowers that attract bees, butterflies, and other beneficial pollinators. Its tender leaves have a mild, warm flavor that is popular in Mediterranean cooking.

NOURISHES:

BEES BUTTERFLIES MOTHS

SUN: Full Sun (6–8+ hours) or Part Shade

HEIGHT: 12–24 in.

GROWING TIP:
Harvest stems regularly to encourage fresh growth, and allow some plants to flower for pollinators.

GARDEN NOTE:
Fresh or dried marjoram adds a mild, sweet flavor to soups, vegetables, poultry, stuffing, and pasta dishes.

Mint

Mentha spp.

A refreshing culinary herb with clusters of small flowers that attract bees, butterflies, and other beneficial pollinators. Its cool, aromatic leaves are widely used in teas, desserts, beverages, and savory dishes.

NOURISHES:

BEES BUTTERFLIES MOTHS

SUN: Full Sun (6–8+ hours) or Part Shade

HEIGHT: 24–48 in.

GROWING TIP:
Plant mint in a container or confined space to prevent it from spreading aggressively.

GARDEN NOTE:
Fresh mint adds bright flavor to teas, fruit, salads, desserts, sauces, and refreshing summer drinks.

Oregano

Origanum vulgare

A flavorful culinary herb with clusters of tiny pink to purple flowers that attract bees, butterflies, and other beneficial pollinators. Its aromatic leaves are a favorite seasoning for Italian and Mediterranean dishes.

NOURISHES:

BEES BUTTERFLIES MOTHS

SUN: Full Sun (6–8+ hours) or Part Shade

HEIGHT: 12–24 in.

GROWING TIP:
Harvest stems regularly to encourage fresh growth, and allow some plants to flower for pollinators.

GARDEN NOTE:
Fresh or dried oregano adds bold flavor to pizza, pasta sauces, roasted vegetables, meats, and Mediterranean recipes.

Parsley

Petroselinum crispum

A versatile culinary herb with clusters of tiny yellow-green flowers that attract many beneficial pollinators. It also serves as an important host for Black Swallowtail butterfly caterpillars.

NOURISHES:

BEES BUTTERFLIES MOTHS

SUN: Full Sun (6–8+ hours) or Part Shade

HEIGHT: 12–24 in.

GROWING TIP:
Harvest outer leaves regularly, and allow second-year plants to flower for pollinators.

GARDEN NOTE:
Fresh parsley brightens soups, salads, sauces, potatoes, vegetables, and countless everyday recipes.

Rosemary

Salvia rosmarinus (formerly Rosmarinus officinalis)

A fragrant herb with pale blue to lavender flowers that attract bees, butterflies, and other beneficial pollinators. Its aromatic leaves are prized for seasoning meats, vegetables, breads, and Mediterranean dishes.

NOURISHES:

BEES BUTTERFLIES MOTHS

SUN: Full Sun (6–8+ hours)

HEIGHT: 24–60 in.

GROWING TIP:
Plant in well-drained soil and avoid overwatering. Prune lightly after flowering to keep plants full and healthy.

GARDEN NOTE:
Fresh or dried rosemary adds bold flavor to roasted potatoes, chicken, lamb, breads, vegetables, and infused oils.

Sage

Salvia officinalis

A fragrant culinary herb with spikes of blue, lavender, pink, or white flowers that attract bees, butterflies, and other beneficial pollinators. Its soft, aromatic leaves are prized for seasoning meats, stuffing, soups, and savory dishes.

NOURISHES:

BEES BUTTERFLIES MOTHS

SUN: Full Sun (6–8+ hours)

HEIGHT: 18–30 in.

GROWING TIP:
Harvest stems regularly to encourage fresh growth, and avoid overwatering to keep plants healthy.

GARDEN NOTE:
Fresh or dried sage adds rich, earthy flavor to stuffing, poultry, sausage, roasted vegetables, and hearty fall recipes.

Summer Savory

Satureja hortensis

A flavorful culinary herb with tiny white to pale lavender flowers that attract bees, butterflies, and other beneficial pollinators. Its peppery, aromatic leaves are popular for seasoning beans, vegetables, meats, and savory dishes.

NOURISHES:

BEES BUTTERFLIES MOTHS

SUN: Full Sun (6–8+ hours)

HEIGHT: 12–18 in.

GROWING TIP:
Harvest stems often to encourage bushy growth, and allow some plants to flower for pollinators.

GARDEN NOTE:
Fresh or dried summer savory adds a mild peppery flavor to beans, soups, vegetables, stuffing, and roasted meats.

Tarragon

Artemisia dracunculus

A flavorful culinary herb with tiny yellow-green flowers that attract bees, butterflies, and other beneficial pollinators. Its slender, aromatic leaves have a delicate anise-like flavor that is prized in French cooking.

NOURISHES:

BEES BUTTERFLIES MOTHS

SUN: Full Sun (6–8+ hours)

HEIGHT: 24–36 in.

GROWING TIP:
Harvest stems regularly to encourage fresh growth, and divide mature plants every few years to keep them vigorous.

GARDEN NOTE:
Fresh or dried tarragon adds delicate flavor to chicken, fish, eggs, vegetables, salad dressings, and herb butter.

Thyme

Thymus vulgaris

A fragrant culinary herb with clusters of tiny pink to lavender flowers that attract bees, butterflies, and other beneficial pollinators. Its small, aromatic leaves are prized for seasoning meats, vegetables, soups, breads, and savory dishes.

NOURISHES:

BEES BUTTERFLIES MOTHS

SUN: Full Sun (6–8+ hours)

HEIGHT: 6–12 in.

GROWING TIP:
Harvest stems often to encourage fresh growth, and avoid overwatering by planting in well-drained soil.

GARDEN NOTE:
Fresh or dried thyme adds earthy flavor to soups, roasted vegetables, poultry, stuffing, breads, and herb blends.

Trees & Shrubs

American Holly

Ilex opaca

American Holly is a beautiful native evergreen tree with glossy, spiny leaves and bright red berries that bring color to the landscape throughout winter. Its tiny spring flowers provide nectar for pollinators, while the berries become an important winter food source for many songbirds.

NOURISHES:

BEES   BUTTERFLIES   MOTHS   SONGBIRDS (SEEDS)

ZONES: 5–9

HEIGHT: 24–48 in.

GARDEN NOTE:
Plant a male and female holly within about 30–40 feet of each other to produce the bright red berries that attract birds and decorate the tree through winter.

American Plum

Prunus americana

American Plum is a hardy native small tree or large shrub covered in fragrant white blossoms each spring. Its early flowers provide an important nectar and pollen source for pollinators, while its sweet summer fruit feeds many songbirds and other wildlife.

NOURISHES:

BEES   BUTTERFLIES   MOTHS   SONGBIRDS (FRUIT)

ZONES: 3–8

HEIGHT: 10–25 ft.

GARDEN NOTE:
American Plum spreads by root suckers to form natural thickets, making it an excellent choice for wildlife hedges, erosion control, and naturalized landscapes.

Bottlebrush

Callistemon spp.

Bottlebrush is an evergreen shrub or small tree known for its vibrant, brush-like red flowers that bloom from spring through summer. Rich in nectar, the blossoms attract a wide variety of pollinators and add bold color to the garden. Its long blooming season makes it one of the best landscape plants for supporting bees, butterflies, and other beneficial wildlife.

NOURISHES:

BEES | BUTTERFLIES | MOTHS | SONGBIRDS (NECTAR)

ZONES: 8–11

HEIGHT: 8–20 ft.

GARDEN NOTE:

Bottlebrush thrives in full sun and is drought tolerant once established. Light pruning after flowering helps maintain its shape and encourages additional blooms.

Buttonbush

Cephalanthus occidentalis

Buttonbush is a beautiful native shrub that produces unique, globe-shaped white flowers throughout summer. Its nectar-rich blooms attract a wide variety of pollinators, while the seeds provide food for many songbirds later in the season. Often found near ponds and streams, it is one of the finest native shrubs for supporting wildlife.

NOURISHES:

BEES | BUTTERFLIES | MOTHS | SONGBIRDS (SEEDS)

ZONES: 5–10

HEIGHT: 6–12 ft.

GARDEN NOTE:

Buttonbush thrives in moist soil and is an excellent choice for ponds, rain gardens, and other wet areas where pollinators gather.

Crabapple

Malus spp.

Crabapple is a small ornamental tree that bursts into colorful blossoms each spring, attracting a wide variety of pollinators. Its small, persistent fruits provide an important food source for many songbirds and other wildlife well into fall and winter.

NOURISHES:

BEES | BUTTERFLIES | MOTHS | SONGBIRDS (FRUIT)

ZONES: 4–8

HEIGHT: 15–25 ft.

GARDEN NOTE:
Choose disease-resistant varieties for the healthiest trees and the most reliable spring blooms and colorful fruit each year.

Dogwood

Cornus florida

Flowering Dogwood is a beautiful native tree known for its showy white or pink spring blooms. Its nectar-rich flowers attract a variety of pollinators, while the bright red berries provide an important food source for many songbirds and other wildlife each fall.

NOURISHES:

BEES | BUTTERFLIES | MOTHS | SONGBIRDS (SEEDS)

ZONES: 5–9

HEIGHT: 15–30 ft.

GARDEN NOTE:
Dogwood grows best in partial shade with well-drained soil. A layer of mulch helps keep the roots cool and moist, especially during hot summers.

Elderberry

Sambucus canadensis

Elderberry is a fast-growing native shrub with large clusters of fragrant white flowers that bloom in early summer. Its nectar-rich blossoms attract a wide variety of pollinators, while the dark purple berries provide an important food source for many songbirds and other wildlife.

NOURISHES:

BEES | BUTTERFLIES | MOTHS | SONGBIRDS (BERRIES)

ZONES: 3–9

HEIGHT: 5–12 ft.

GARDEN NOTE:
Elderberry grows best in full sun to partial shade with moist, well-drained soil. Plant two or more shrubs nearby for the best fruit production.

Fringe Tree

Chionanthus virginicus

Fringe Tree is a beautiful native tree covered with delicate, fragrant white flowers in late spring. Its nectar-rich blooms attract a variety of pollinators, while its dark blue fruits provide food for many songbirds and other wildlife in late summer.

NOURISHES:

BEES | BUTTERFLIES | MOTHS | SONGBIRDS (FRUIT)

ZONES: 3–9

HEIGHT: 12–20 ft.

GARDEN NOTE:
Fringe Tree grows best in full sun to partial shade with well-drained soil. Its graceful spring flowers make it an excellent specimen tree for smaller landscapes.

Magnolia

Magnolia grandiflora

Southern Magnolia is a majestic evergreen tree known for its large, fragrant white flowers and glossy green leaves. Its nectar-rich blooms attract a variety of pollinators, while the bright red seeds provide food for many songbirds and other wildlife in fall.

NOURISHES:

BEES BUTTERFLIES MOTHS SONGBIRDS (SEEDS)

ZONES: 4–9

HEIGHT: 20–30 ft.

GARDEN NOTE:
Southern Magnolia grows best in full sun to partial shade with rich, well-drained soil. Give it plenty of room to mature, as it develops into a large, long-lived shade tree.

Redbud

Cercis canadensis

Eastern Redbud is a beautiful native tree that bursts into rosy-pink blossoms in early spring before its heart-shaped leaves emerge. Its nectar-rich flowers provide an important early food source for pollinators, while the seeds feed many songbirds and other wildlife later in the season.

NOURISHES:

BEES BUTTERFLIES MOTHS SONGBIRDS (SEEDS)

ZONES: 4–9

HEIGHT: 20–30 ft.

GARDEN NOTE:
Redbud grows best in full sun to partial shade with well-drained soil. Its early spring blooms make it one of the first trees to brighten the garden after winter.

Serviceberry

Amelanchier spp.

Serviceberry is a beautiful native small tree covered with delicate white blossoms in early spring. Its nectar-rich flowers attract a variety of pollinators, while its sweet purple berries provide food for many songbirds and other wildlife in early summer.

NOURISHES:

BEES | BUTTERFLIES | MOTHS | SONGBIRDS (BERRIES)

ZONES: 4–9

HEIGHT: 15–25 ft.

GARDEN NOTE:

Serviceberry grows best in full sun to partial shade with well-drained soil. Its beautiful spring flowers, colorful fall foliage, and edible berries make it a favorite for both gardens and wildlife.

Sourwood

Oxydendrum arboreum

Sourwood is a beautiful native tree with graceful, drooping clusters of fragrant white flowers that bloom in midsummer. Its nectar-rich blossoms attract a wide variety of pollinators, while the small seeds provide food for many songbirds and other wildlife.

NOURISHES:

BEES | BUTTERFLIES | MOTHS | SONGBIRDS (SEEDS)

ZONES: 5–9

HEIGHT: 25–40 ft.

GARDEN NOTE:

Sourwood grows best in full sun to partial shade with moist, well-drained, acidic soil. Its brilliant fall foliage adds spectacular color to the landscape.

Spicebush

Lindera benzoin

Spicebush is a fragrant native shrub covered with tiny yellow flowers in early spring before its leaves emerge. Its nectar-rich blooms attract pollinators, while the bright red berries provide food for many songbirds and other wildlife in fall.

NOURISHES:

BEES | BUTTERFLIES | MOTHS | SONGBIRDS (BERRIES)

ZONES: 4–9

HEIGHT: 6–12 ft.

GARDEN NOTE:
Spicebush grows best in partial shade with moist, well-drained soil. Its aromatic leaves, colorful berries, and brilliant fall foliage make it an excellent addition to native gardens.

Summersweet

Clethra alnifolia

Summersweet is a fragrant native shrub covered with spikes of fragrant flowers that bloom in midsummer. Its nectar-rich blossoms attract a wide variety of pollinators, while the small seeds provide food for many songbirds and other wildlife later in the season.

NOURISHES:

BEES | BUTTERFLIES | MOTHS | SONGBIRDS (SEEDS)

ZONES: 4–9

HEIGHT: 3–8 ft.

GARDEN NOTE:
Summersweet grows best in full sun to partial shade with moist, well-drained soil. Its fragrant flowers and long bloom season make it an excellent choice for pollinator and rain gardens.

Sweetbay Magnolia

Magnolia virginiana

Sweetbay Magnolia is a graceful native tree with fragrant, creamy white flowers that bloom from late spring into summer. Its nectar-rich blossoms attract a variety of pollinators, while its bright red seeds provide food for many songbirds and other wildlife in fall.

NOURISHES:

BEES · BUTTERFLIES · MOTHS · SONGBIRDS (SEEDS)

ZONES: 5–10

HEIGHT: 15–30 ft.

GARDEN NOTE:
Sweetbay Magnolia grows best in full sun to partial shade with moist, slightly acidic soil. Its fragrant flowers and silvery leaf undersides add beauty to gardens throughout the growing season.

Viburnum

Viburnum spp.

Viburnum is a versatile shrub with clusters of fragrant white flowers that bloom from spring into early summer. Its nectar-rich blossoms attract a variety of pollinators, while its colorful berries provide food for many songbirds and other wildlife in fall.

NOURISHES:

BEES · BUTTERFLIES · MOTHS · SONGBIRDS (BERRIES)

ZONES: 3–9

HEIGHT: 6–15 ft.

GARDEN NOTE:
Viburnum grows best in full sun to partial shade with well-drained soil. Plant two compatible varieties nearby for the best berry production and a more abundant wildlife display.

INDEX

www.ingramcontent.com/pod-product-compliance
Lightning Source LLC
LaVergne TN
LVHW051157210926
848486LV00016B/400/J